DINOSAURS AND THEIR DISCOVERERS™

The Dinosaur Bone Battle Between O.C. Marsh and Edward Drinker Cope

Brooke Hartzog

The Rosen Publishing Group's
PowerKids Press™
New York

Published in 1999 by The Rosen Publishing Group, Inc.
29 East 21st Street, New York, NY 10010

First Edition

Book Design: Danielle Primiceri

Photo Credits: p. 5 © Tom Bean/Corbis; pp. 6, 8 © American Museum of Natural History; pp. 7, 9, 10, 13, 14, 18, 20, 21, 22 © Linda Hall Library.

Hartzog, Brooke.
   The dinosaur bone battle between O.C. Marsh and Edward Drinker Cope / by Brooke Hartzog.
      p.   cm.—(Dinosaurs and their discoverers)
   Summary: Tells the story of two nineteenth-century paleontologists who used questionable tactics as they tried to outdo each other in collecting dinosaur bones.
   ISBN 0-8239-5327-0
   1. Cope, E.D. (Edward Drinker), 1840–1897—Juvenile literature. 2. Marsh, Othneil Charles, 1831–1899—Juvenile literature. 3. Paleontologists—United States—Biography—Juvenile literature. 4. Paleontology—United States—History—19th century—Juvenile literature. [1. Cope, E. D. (Edward Drinker), 1840–1897. 2. Marsh, Othneil Charles, 1831–1899. 3. Paleontologists. 4. Paleontology.] I. Title. II. Series: Hartzog, Brooke. Dinosaurs and their discoverers.
   QE22.C56H37 1998
   560'.973—dc21
                                                          98-10336
                                                           CIP
                                                           AC

Manufactured in the United States of America

Contents

The Battle for Bones

In 1877 a schoolteacher named Arthur Lakes sent a letter that started a **battle** (BA-tul). It wasn't a battle between countries over land. It was a battle over dinosaur bones. The people battling weren't soldiers. They were **paleontologists** (pay-lee-un-TAH-luh-jists). In the letter, Arthur told two important paleontologists about his discovery of some dinosaur bones in Colorado. The paleontologists didn't want to work together to find more bones. Instead, each one wanted to find the most dinosaur bones and keep the other from finding any.

Paleontologists study the fossil remains of plants and animals that lived long ago.

O. C. Marsh

O. C. Marsh was a paleontologist at Yale University. He was a handsome man with a beard. He never got married because he loved **fossils** (FAH-sulz) more than anything else. And he worked all the time to try to find them. His uncle had given him a large amount of money to start a **museum** (myoo-ZEE-um). The museum had fossils of **extinct** (ek-STINKT) plants and animals. But it didn't have any dinosaur fossils.

These are the fossilized remains of a dinosaur called a stegosaurus.

Edward Drinker Cope

When Marsh got the letter about dinosaur bones in Colorado, he was excited. He would finally be able to get some dinosaur **skeletons** (SKE-luh-tunz) for his museum. He wanted his museum to have more dinosaur fossils than any other museum. Like O. C. Marsh, Edward Drinker Cope was a paleontologist.

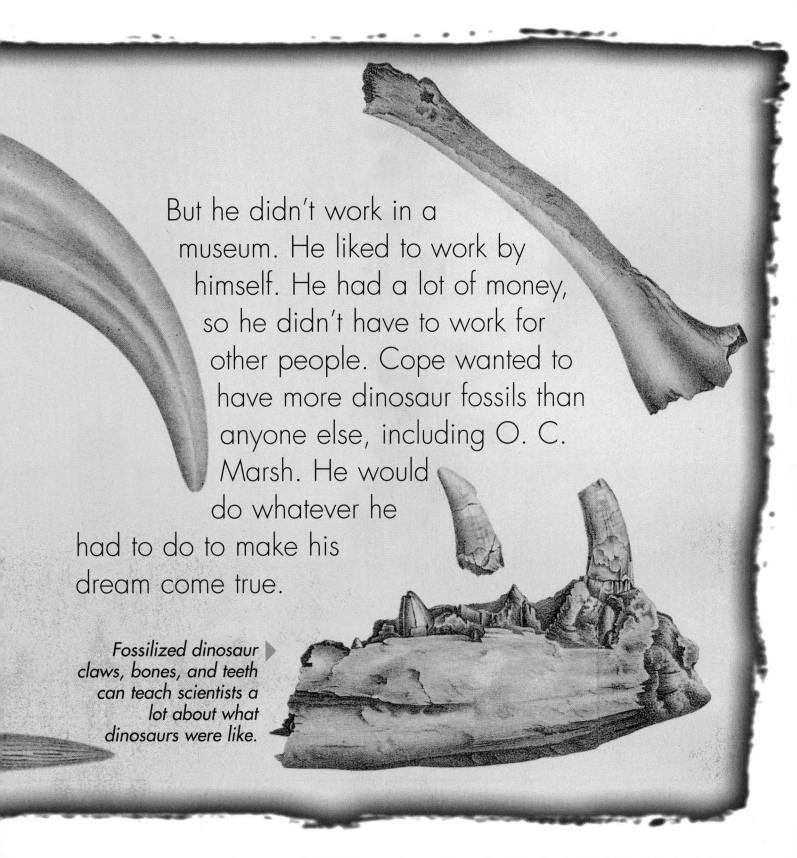

But he didn't work in a museum. He liked to work by himself. He had a lot of money, so he didn't have to work for other people. Cope wanted to have more dinosaur fossils than anyone else, including O. C. Marsh. He would do whatever he had to do to make his dream come true.

Fossilized dinosaur claws, bones, and teeth can teach scientists a lot about what dinosaurs were like.

Marsh Buys Fossils

Marsh and Cope found out that they had both been sent letters about the Colorado fossils. Each wanted to get the fossils first. Marsh sent one hundred dollars to the schoolteacher who had found the big bones. One hundred dollars was a lot of money in 1877. The schoolteacher decided to send all his fossils to Marsh. Cope was very angry about this.

It seemed Marsh had won the first battle of the Bone Wars. Cope refused to give up. He **vowed** (VOWD) he wouldn't let Marsh get any more dinosaur bones.

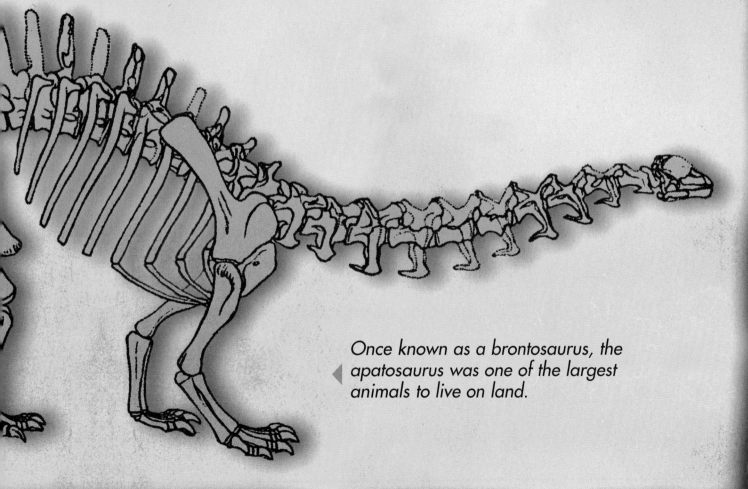

Once known as a brontosaurus, the apatosaurus was one of the largest animals to live on land.

Cope's Secret Site

Cope was lucky enough to get a letter from another schoolteacher, O.W. Lucas, who had also found some giant dinosaur bones. These fossils turned out to be bones from a dinosaur called a camarasaurus. Cope tried to keep this new find a secret. Somehow Marsh found out about the fossils. But men who worked for Cope had already taken the best fossils in the area. Now it seemed like Cope was winning the battle. Marsh went back to the place where the first bones had been found. One day a huge rock slide killed everyone who was digging up fossils that day. Marsh soon had to **abandon** (uh-BAN-dun) that **site** (SYT).

Camarasaurus was a plant-eating dinosaur found in North America. ▶

Fossils for Sale

In July 1877 Marsh received a **mysterious** (mih-STEER-ee-us) letter from Laramie, Wyoming. The letter was signed by two men, named Harlow and Edwards. The letter said the two men had found some huge fossils. They also said they thought there were many more dinosaur bones in the area. They kept the **location** (loh-KAY-shun) of the fossils a secret. The men told Marsh they would sell him the fossils. Harlow and Edwards also said that they would look for more fossils if Marsh sent them some money.

The plant-eating apatosaurus was over 70 feet long and weighed 27 tons.

Secrets of the Desert

Marsh started paying the two men to hunt for more fossils. He then found out the names the men signed on the letter were fake. Their real names were W. E. Carlin and W. H. Reed. They worked on the railroad and had discovered the fossils on Como Bluff, Wyoming, while hunting for antelope. Marsh made them promise not to tell anyone about the fossils. He also made sure the men sold fossils only to him. He wanted the fossils for his museum in Connecticut. And he didn't want Cope to get any of the fossils.

Cope and Marsh discovered important fossil remains in Colorado and Wyoming. The fossils were usually shipped to museums in Connecticut and New York.

New York

Massachusetts

CONNECTICUT

New Jersey

United States
of America

The News Spreads

It was hard to keep a **dig** (DIG) secret in the small town in Wyoming. Soon the local newspaper wrote a story about the fossil hunters. Marsh was worried. He sent his assistant, Samuel Williston, to check on the **excavation** (eks-kuh-VAY-shun). Williston noticed a strange man hanging around the site. He was a fat man who walked with a limp. And he had a frown on his face. The man said his name was Haines and he was there to sell groceries. Williston knew right away that this man was really a spy for Cope.

This gives you an inside and outside look at a claosaurus.

Cope Invades

Soon after the spy was discovered, Marsh's **territory** (TEHR-uh-tohr-ee) was invaded by men working for Cope. Marsh didn't own the land where his men had found the dinosaur fossils. So he couldn't stop Cope from fossil hunting there. But Marsh and Cope were **enemies** (EH-nuh-meez). After Marsh took as many fossils as he could from the site, he told his workers to smash the rest of the fossils so that Cope wouldn't get any.

A diplodocus's neck had fifteen bones that were light but strong.

Famous Fossils

O. C. Marsh and Edward Drinker Cope were both very selfish. They each wanted to be the most important paleontologist in the world. And they each wanted the other to fail. Their dinosaur bone battles produced some of the most important fossil discoveries in American history. Almost every natural history museum in America has at least one skeleton found by Cope or Marsh. Just think what they could have done if they had helped each other instead of fighting with each other.

Web Sites:

You can learn more about dinosaurs at this Web site: www.questionmark.com.au/qm_web/dino.html

Glossary

abandon (uh-BAN-dun) To leave a place and plan to never come back.

battle (BA-tul) One of several fights between two people.

dig (DIG) A search for dinosaur bones.

enemy (EH-nuh-mee) A person who works against you.

excavation (eks-kuh-VAY-shun) Digging up something that was buried in the ground or covered by rocks.

extinct (ek-STINKT) To no longer exist.

fossil (FAH-sul) The hardened remains of a dead animal or plant.

location (loh-KAY-shun) Where something is taking place.

museum (myoo-ZEE-um) A building where pieces of art or historical items are displayed.

mysterious (mih-STEER-ee-us) Strange and hard to understand.

paleontologist (pay-lee-un-TAH-luh-jist) Someone who studies things that were alive in the past, such as dinosaurs.

site (SYT) The scene of a certain event.

skeleton (SKE-luh-tun) The set of all the bones in an animal's body.

territory (TEHR-uh-tohr-ee) Land that is controlled by a person or a group of people.

vow (VOW) To make a very important promise.

Index